The ABC's of Women Inventors, Scientists, & Engineers

Sandra Kolvick, P.E.

Jill S. Tietjen, P.E.

Graphic Design by Ashley Wynne, Ph.D.

Image Credit: A: Grace Murray Hopper, At Work (Photograph courtesy of Society of Women Engineers Records, Walter P. Reuther Library, Wayne State University)
All other images are public domain or the creation of the illustrator

Library of Congress Cataloging-in-Publication Data

Names: Kolvick, Sandra, author. | Tietjen, Jill S., author. | Wynne, Ashley, illustrator.
Title: The ABC's of Women Inventors, Scientists, & Engineers / Sandra Kolvick, PE, Jill S. Tietjen, PE., Ashley Wynne, PhD
Description: Greenville, South Carolina
Identifiers: ISBN 978-0-9861791-3-6
LCCN Library of Congress Control Number: 2025923644

Printed in the United States
0 9 8 7 6 5 4 3 2 1

A is for **Admiral Grace Murray Hopper**,

Mathematician and Physicist

Grace invented the computer compiler that allows us to talk to a computer with regular languages. She also found the first computer bug – a moth! It had gotten stuck in the computer. She removed it and taped it in her notebook.

"The most dangerous phrase in the language is, 'We've always done it this way.'"

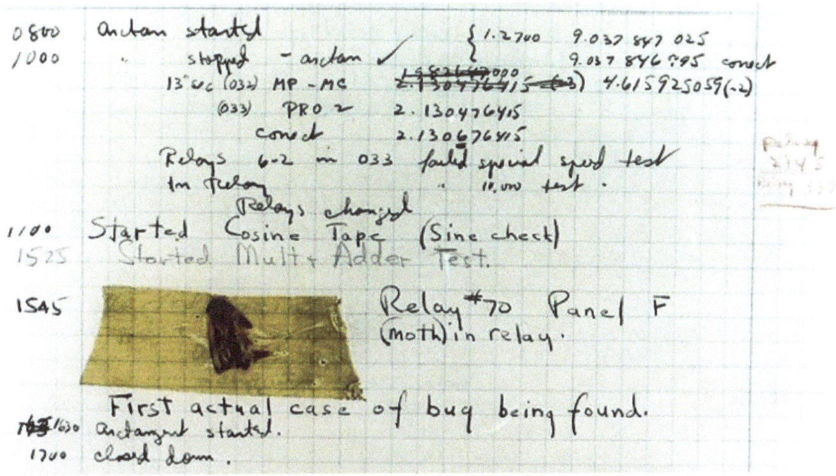

B is for Beatrice Hicks,

Chemical Engineer

Beatrice (Bea) invented a sensor for gas density that could work in outer space! Her device was used on rockets that took people to the moon.

Bea was the co-founder and 1st President of the Society of Women Engineers in 1950.

Bea's Patent:

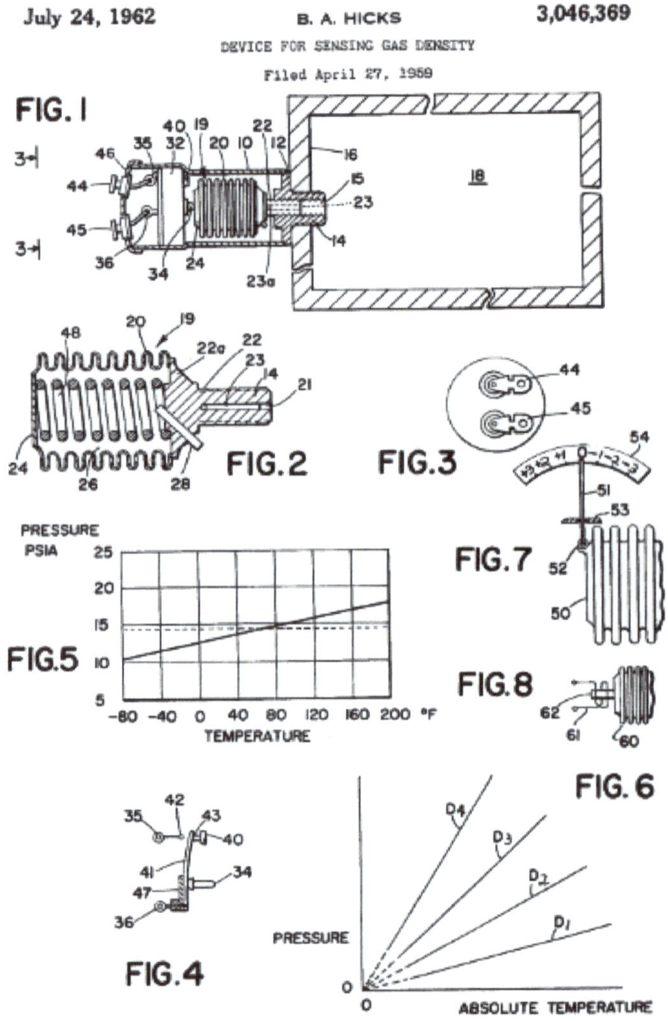

"...education and work in engineering is both the responsibility and the privilege of all whose talents are scientific in nature."

C is for Chieko Asakawa,

Japanese Computer Scientist

Chieko created the most popular way for blind people to use the internet. Her application reads aloud the text and describes the icons, and a person can navigate by using the numbers on their keyboard!

"I had an idea that a computer could help to bridge the gap between sighted and blind people ... I thought I could effectively use my own experience as a blind person working in science and technology to make it come true."

Her 2 daughters were 10 and 15 when she earned her PhD in computer science

D is for **D**octor Virginia Apgar,

Anesthesiologist and Physician

Virginia invented an infant health test in 1952, that is now used in most hospitals worldwide. Chances are **you** had this test at one and five minutes after you were born. The **APGAR** Score is used to determine what type of medical care a baby might need right away. Each letter gets a 0, 1, or 2.

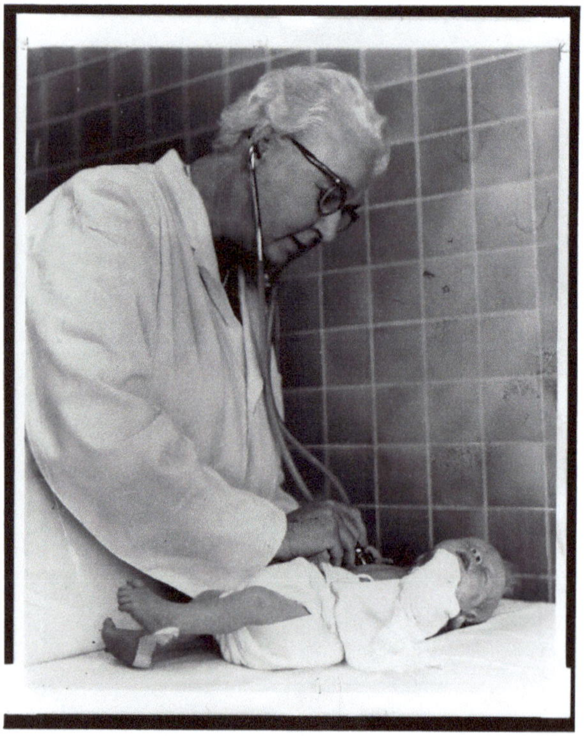

"I kept wondering who was really responsible for the newborn. Birth is the most hazardous time of life. It is urgently important to evaluate quickly the status of a just-born baby and to identify immediately those who need emergency care."

A - **A**ppearance
P - **P**ulse
G - **G**rimace
A - **A**ctivity
R - **R**espiration

E is for **Edith Clarke**,

Electrical Engineer

Edith invented the **Clarke Calculator**, a faster way to solve equations for the wires that carry electricity to where you live. Her math made it easier for us to plan modern electric grids.

Edith's Patent:

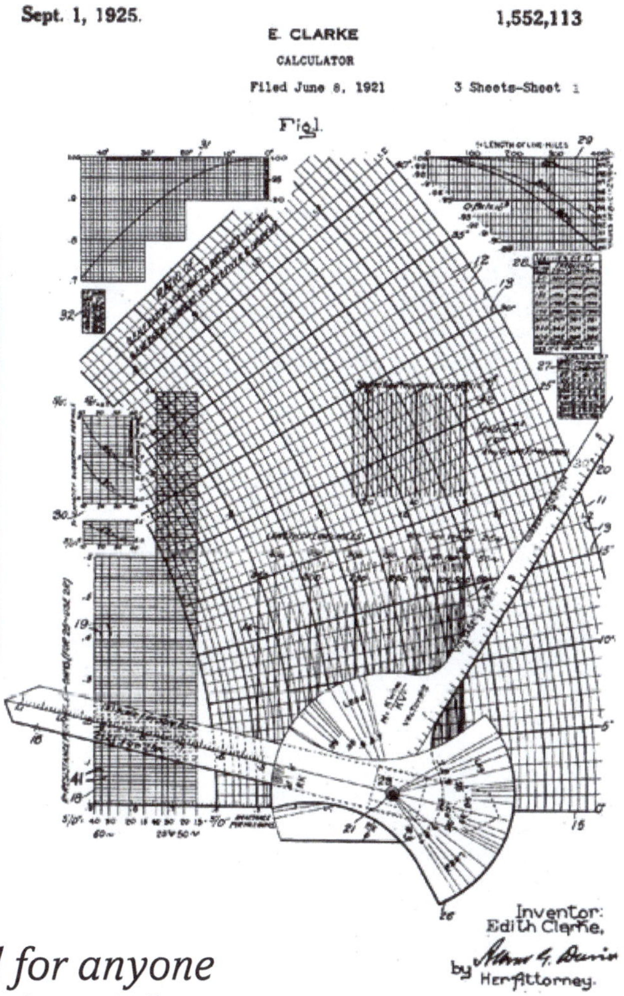

"There's always a demand for anyone who can do a good piece of work."

F is for **Flossie Wong-Staal,**

Chinese-American Molecular Biologist

Flossie worked with viruses and was the first to complete genetic mapping of the virus called HIV (Human Immunodeficiency Virus). This made it possible to develop tests for infections. Flossie's research used gene therapy to target the virus in **just** the infected cells.

"Working with this virus is like putting your hand in a treasure chest. Every time you put your hand in, you pull out a gem."

Flossie was the mother of 2 daughters

G is for Gertrude Elion,

Biochemist

Gertrude (Trudy) developed new ways to make drugs by building new molecules called purines. One of her drugs was the first successful drug to treat childhood leukemia. Another drug she developed made organ transplants possible.

"It's amazing how much you can accomplish when you don't care who gets the credit."

H is for **H**elen Greiner,

British-American Mechanical Engineer 

Helen invented the Roomba robot vacuum and co-founded the company that makes them. Her other robots have explored the surface of Mars and inside pyramids in Egypt.

"I see more and more women becoming entrepreneurs."

I is for Irène Joliot-Curie,

French Mathematician and Physicist

Irene discovered artificial radioactivity. She would take stable elements and **smack** them with alpha particles! She made elements radioactive so that they can be used for medical treatments and for making nuclear electricity.

"The farther the experiment is from theory, the closer it is to the Nobel Prize."

Irène was the mother of Hélène, a nuclear physicist and Pierre, a biochemist

J is for Julia Morgan,

Civil Engineer and Architect

Julia was the first to use reinforced concrete in her buildings that withstood earthquakes. As the first female licensed architect in California, she designed 700 buildings in her career, including churches and castles!

Hearst Castle, San Simeon, CA is now a museum!

"My buildings will be my legacy... they will speak for me long after I'm gone."

K is for Kristina Johnson,

Electrical Engineer

When you see a 3D movie, you are using Kristina's optoelectronics technology that splits a ray of light in half. Her special film and glasses let us see in 3D!

16th president of The Ohio State University

"Optimism, kindness, and community are all essential ingredients for a flavorful life."

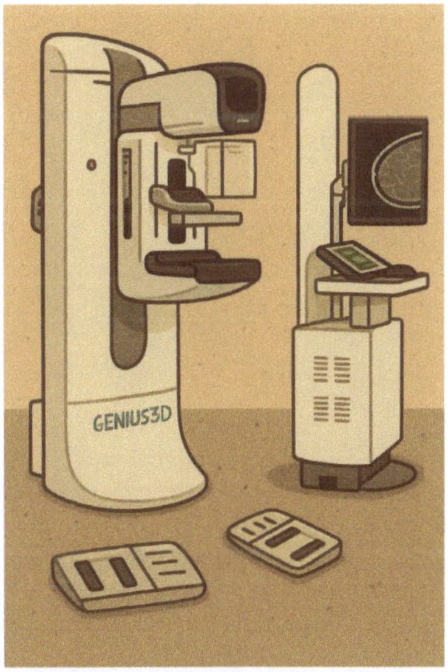

3D digital mammogram machine with Kristina's technology

L is for **Lillian Moller Gilbreth**,

Industrial Engineer

Lillian founded the field of industrial engineering, which is how to do something the most efficient way. She had 12 children so that she and her husband could test their theories. She designed many of the drawers and features inside a refrigerator. She also designed and patented the step that opens the lid on a trashcan!

Lillian earned her PhD from Brown University in 1915

Hollywood made movies on her life: *Cheaper by the Dozen* (1950) and *Belles on Their Toes* (1952)

"To meet today's challenge to women, one must make up her own philosophy of life."

M is for **Mary** Engle Pennington,

Microbiologist and Refrigeration Engineer

Mary rode on top of railroad refrigeration cars all over the country to test their efficiency. She invented new cooling systems and redesigned the railcars to keep our food safe for thousands of miles! She knew this was essential for safe eggs, fish, chicken, and milk.

One of Mary's 6 Patents:

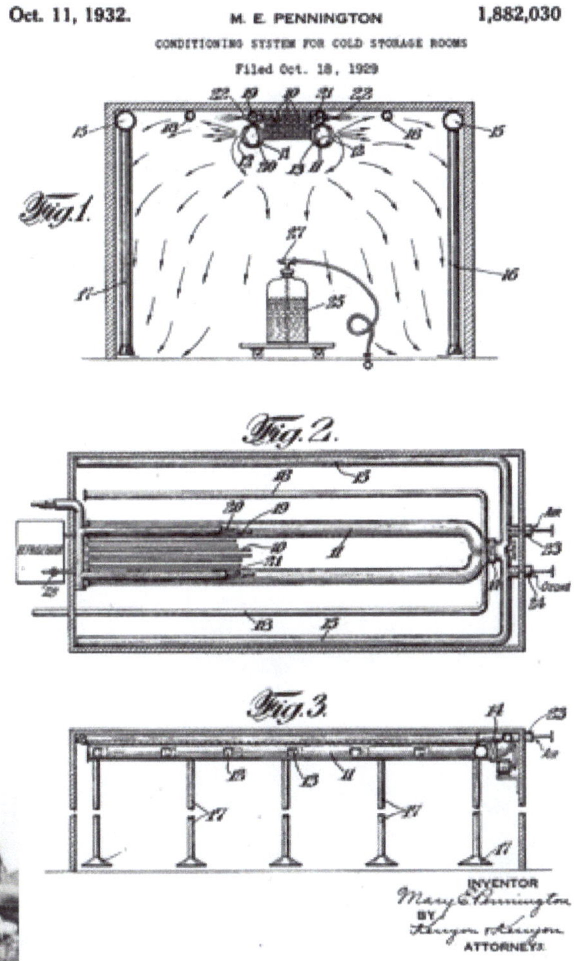

"There is a thrill when a scientific idea suddenly strikes home in the form of a practical solution to an industrial problem."

N is for **Nystatin**, the first antifungal medicine invented by Elizabeth Lee Hazen, microbiologist and Rachel Fuller Brown, chemist.

Elizabeth and Rachel analyzed **dirt!** They found a bacteria that could be used to treat fungus diseases. They named it for the New York State Department of Health. It was the first safe antifungal medicine for humans, curing infections and athlete's foot. It was even used to stop mold on historic art in Italy!

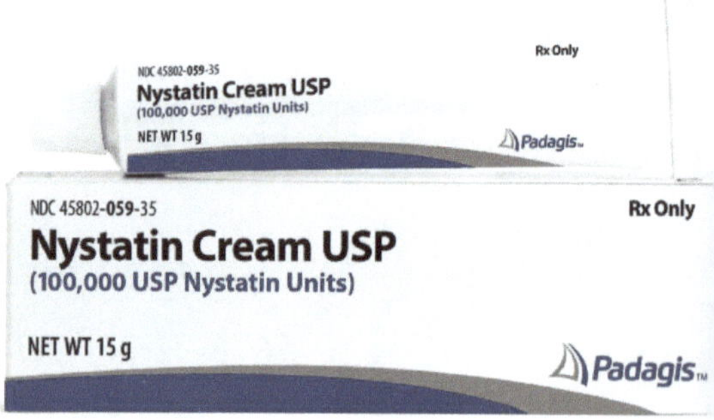

Hazen and Brown donated their royalties, more than $13 million, to academic science.

O is for **Ophthalmologist**

Patricia Bath

Patricia specialized in laser cataract surgery. She invented a device that uses lasers to dissolve cataracts. Her device is a much better and safer way to remove cataracts and is widely used today.

"I had a few obstacles but I had to shake it off"

Patricia's Patent:

U.S. Patent **May 17, 1988** **4,744,360**

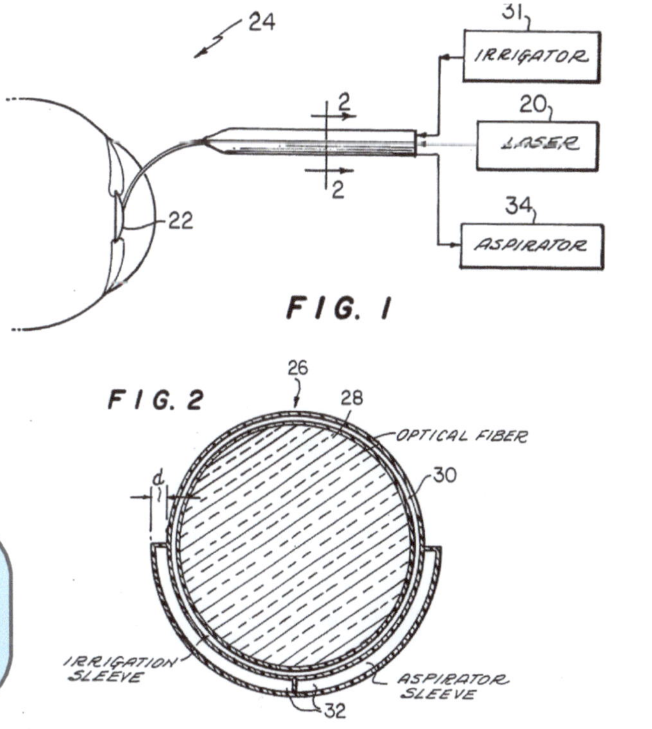

Her daughter Eraka Bath said: "She wore sneakers and jeans... she was casual and not pretentious."

P is for **P**aramjit Khurana,

Indian Scientist and Professor in Plant Biotechnology, Molecular Biology, and Genomics

Paramjit has used genetic engineering to make plants such as mulberry, wheat, and rice easier to grow and harvest. Her plants survive better in heavy stress, high heat, and strong sunlight.

"Ten years down the line, I will still be working in my laboratory. I hope my products benefit everyone."

Q is for the **Queen of Carbon Science**

Millie Dresselhaus, Physicist

Millie was known as the Queen of Carbon Science because she worked to understand carbon at a molecular level. She made materials such as graphene, which is harder than a diamond and 10 times stronger than steel! This semi-metal is used in energy, construction, healthcare, and electronics.

"I like to be challenged. I welcome the hard questions and having to come up with good explanations on the spot. That's an experience I really enjoy."

Millie was the mother of 4 children

R is for **Ruby Payne-Scott**,

Australian Physicist

Ruby worked in the field of radio astronomy. She noticed that radio emissions were coming from the sun in 'bursts' from sunspots. She developed a machine to record sunspot radio emissions every second.

"We're climbing up on ladders, up on aerials every day.

I'm not going up on a ladder with a skirt on."

Ruby was the mother of Peter, a mathematician, and Fiona, an artist.

S is for **Shirley Jackson,**

Physicist

Shirley developed a math model for materials that could send light over long distances, the foundation of fiber optic cables! These cables, made of glass or plastic, are used for high-speed data transmission.

18th president of Rensselaer Polytechnic Institute in New York

Fiber Optic Cable

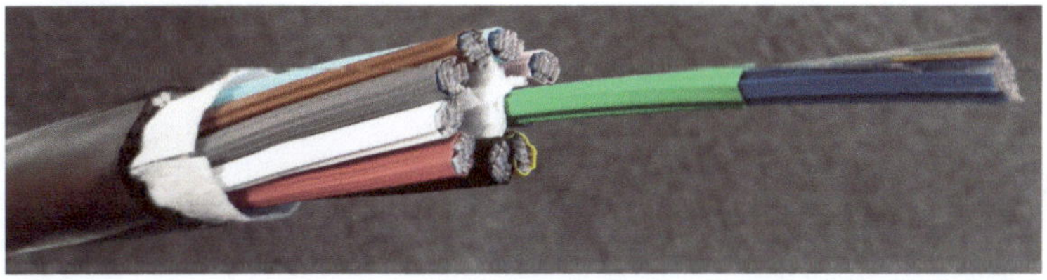

"If you're at a standstill, keep working. Important things tend to be hard."

Shirley is the mother of Alan Washington

T is for Tu Youyou,

Chinese Pharmaceutical Chemist 🇨🇳

Tu developed a new treatment for malaria that saved millions of lives. The drug artemisinin (qīnghāosù, 青蒿素).was because of her! She read and analyzed traditional Chinese medical books, then combined that knowledge with a new low temperature process to make the drug effective.

On receiving the Nobel Prize, she said:

"Every scientist dreams of doing something that can help the world."

U is for USGS, and Florence Bascom,

Geologist at the **U**nited **S**tates **G**eological **S**urvey

Florence was the first professional female geologist hired by the USGS in 1896. She was the first person to survey Mount Desert Island in Acadia, Maine and published her report in 1919. She was called *The Stone Lady*.

Florence's drawings of volcanic rock formations:

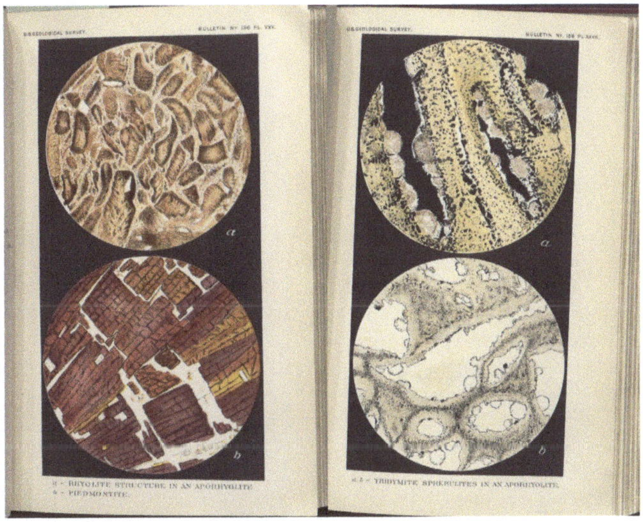

"When any woman manifests an interest in the science [of geology] I am always glad to tell her of its possibilities and she makes her own choice."

Florence's Poem : The Hut

A hut and a tree
And a hill for me!
A dog, a horse, and upland meadows,
And I'll ask no thing
Of man or king
Save to clear away their shadows.

V is for Valerie L. Thomas,

Data Scientist and Physicist

Valerie invented and patented a device called the illusion transmitter in 1980. It uses mirrors to transmit 3D images. NASA uses this technology in satellites; on Earth it is used in surgery and 3D TV.

Valerie's Patent:

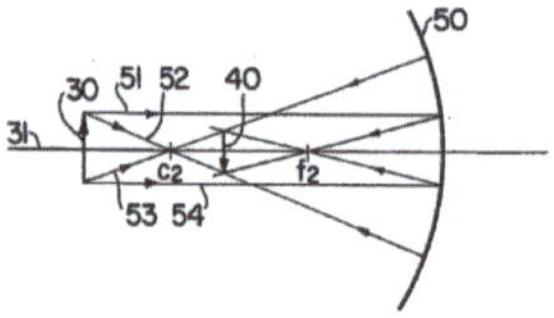

Fig. 2

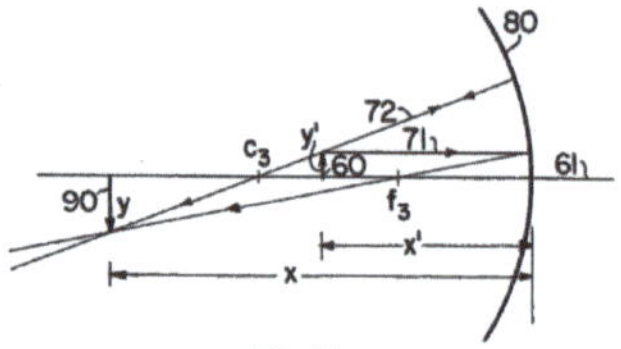

Fig. 3

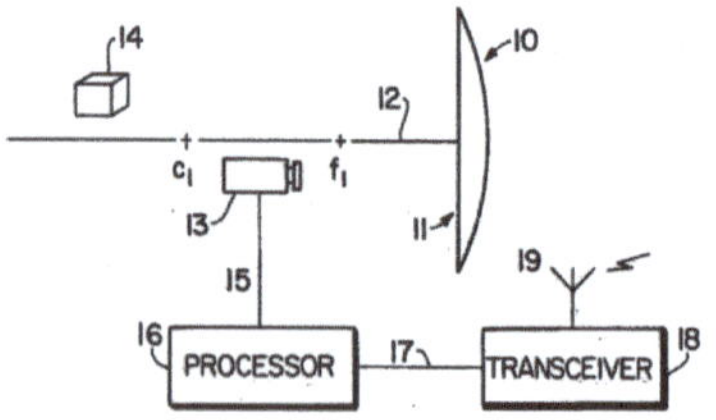

Fig. 1(a)

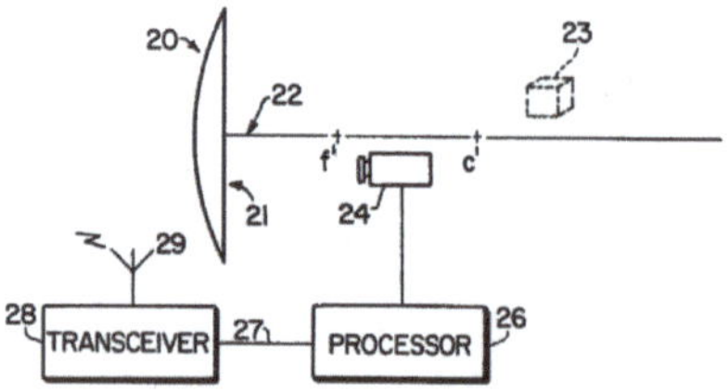

Fig. 1(b)

"My purpose in life is to help people and cause positive things to happen."

W is for the **Walnut Queen**, Harriet Strong, Water Conservationist

Harriet was an innovator in dry irrigation, water storage, and flood control. Her designs helped make California a giant in agriculture. Her ranch led the country in walnut production, earning her the nickname of "Walnut Queen".

"I'd train every girl so that... she would rely upon her own judgment."

Harriet was the mother of Mary, Harriet, Nelle, & Georgina

One of Harriet's 5 patents:

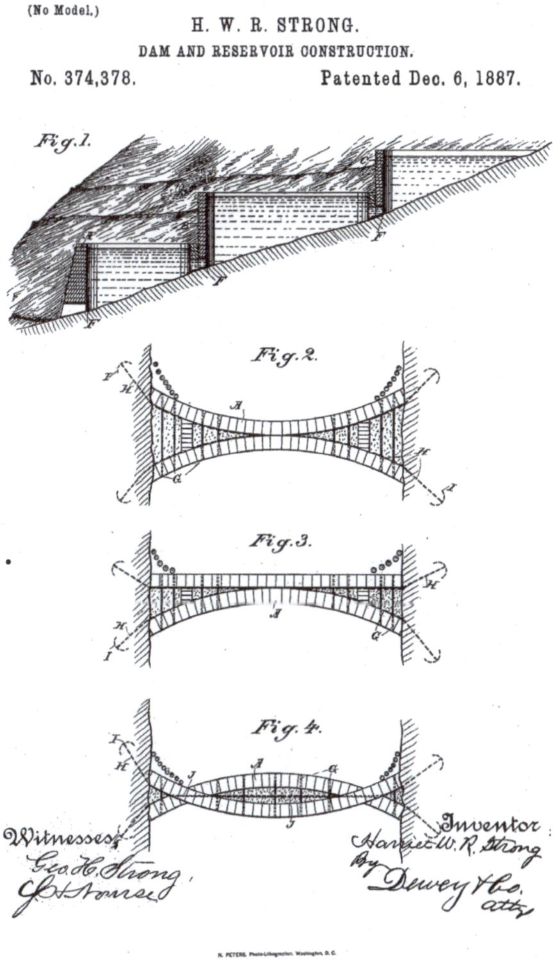

X is for X-rays, and Marie Curie,

Polish-French Physicist and Chemist

Marie developed mobile radiology with x-rays. She also discovered the elements Radium and Polonium. She is the only woman with two Nobel Prizes!

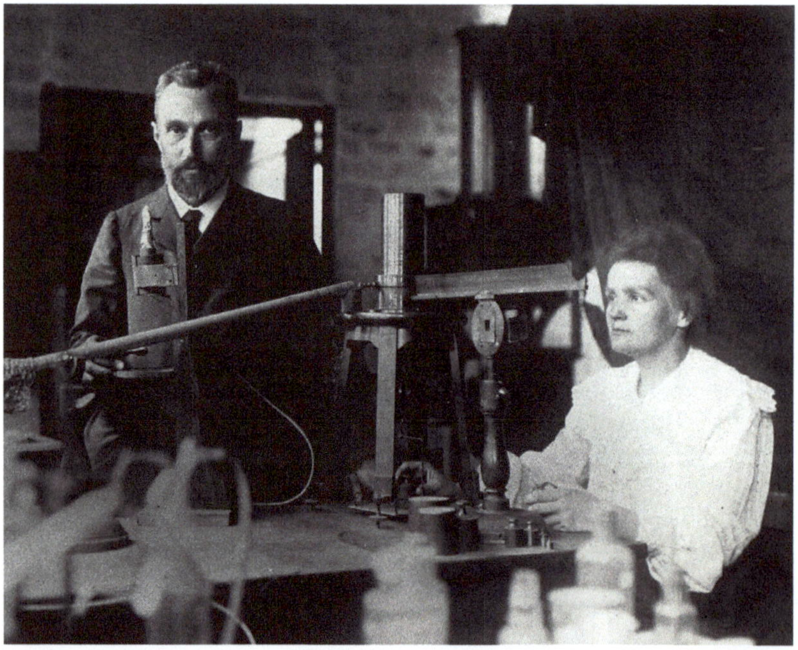

Marie was the mother of Irène and Ève

"Life is not easy for any of us. But what of that? We must have perseverance and above all confidence in ourselves. We must believe that we are gifted for something, and that this thing, at whatever cost, must be attained."

Y is for Yvonne Brill,

Canadian-American Jet Propulsion Engineer

Yvonne worked in satellite propulsion and invented the **resistojet**. It uses electricity to carefully heat the fuel that expands and pushes the satellite forward. Nearly all satellites today can stay in orbit longer with this much more precise and efficient technology.

“You just have to be cheerful about it and not get upset when you get insulted.”

[Yvonne had a telescope mounted by the garage.] "the kids used to climb up, and we could see the rings of Saturn."

Z is for **Z**oologist Rachel Carson,

Rachel showed the dangers of the pesticide DDT, such as birds laying eggs with thinner shells, by writing a book called *Silent Spring*. The chemical was banned so it would no longer be in our food.

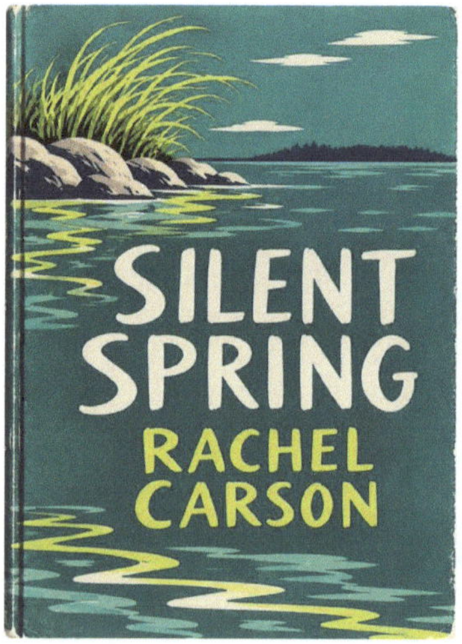

"All the people of a country have a direct interest in conservation . . . Wildlife, water, forests, grasslands – all are part of man's essential environment; the conservation and effective use of one is impossible except as the others are also conserved."

About the Authors & Graphic Designer:

Sandra Kolvick, P.E. is a mechanical engineer. She was dissatisfied with children's books that did not explain what technical women actually did. She is a licensed Professional Engineer, a SWE and ASME life member, and holds multiple patents.

Jill S. Tietjen, P.E. is a best-selling and award-winning author with 19 books published through 2025. An electrical engineer in the electric utility industry, she is a Fellow and Past President of SWE. She is a Lifetime Girl Scout.

Ashley Wynne, Ph.D. is a chemical engineer in research and product development and a recognized leader and mentor who champions equity in STEM through professional society service and university engagement.

We connected through the Society of Women Engineers (SWE).

Additional Enrichment

National Inventors Hall of Fame: www.invent.org
National Women's Hall of Fame: www.womenofthehall.org

Additional Publications by Jill S. Tietjen:

Her Story: A Timeline of the Women Who Changed America www.herstoryatimeline.com
Hollywood: Her Story, An Illustrated History of Women and the Movies www.hollywoodherstory.com
Over, Under, Around, and Through: How Hall of Famers Surmount Obstacles https://bit.ly/OUATBook
Duty Calls: Lessons Learned from an Unexpected Life of Service https://bit.ly/DutyCalls-ALifeofService

www.ingramcontent.com/pod-product-compliance
Lightning Source LLC
LaVergne TN
LVHW070627061126
851857LV00002B/25
* 9 7 8 0 9 8 6 1 7 9 1 3 6 *